Fact File
MAMMALS

Author: Kylie Currey
Principal photographer: Steve Parish

Introduction

WHAT MAKES A MAMMAL A MAMMAL?

One of the main features that sets a mammal apart from all other animals is that a mammal is covered in fur and feeds its babies milk. Humans are mammals because they are covered in fur, which is another name for hair. Mammals are "vertebrates", which means that they have a skeleton that is supported by a backbone. They are warm-blooded and give birth to live young. A female mammal makes milk in her *mammary glands,* and her young suckle, or drink the milk from her *teats.*

MAMMAL SPECIES

There are close to 268 different types of mammal living in Australia and more than 80 percent of these are found nowhere else on the planet! Each type of mammal is a different *species.* Each species has its own special two-word Latin name, also known as its scientific name, which is different from the name given to every other animal in the world. Giving each species of animal a scientific name makes sure there is no confusion when we talk about different animals using their common names. For example, there are two species of echidnas in the world — one species lives only in Australia, while the other lives only in Papua New Guinea. The Australian echidna has the scientific name, *Tachyglossus aculeatus,* while the echidna that lives in Papua New Guinea has the scientific name, *Zaglossus bruijni.* If we were to talk about one of them using only its common name, echidna, how would we know exactly which species we were talking about? By using the Latin name, we make sure there is no confusion. A species' scientific name also allows anybody, no matter where they live in the world or what language they speak, to understand exactly which species of animal we are talking about.

The red kangaroo is Australia's largest marsupial.

The eastern pygmy-possum is one of Australia's smallest mammals.

THREE GROUPS OF MAMMALS

Each different species of mammal in Australia is placed into one of three groups — monotremes, marsupials and placental mammals — depending on how they give birth to their young.

Monotremes

Monotremes are different from other mammals because they lay soft-shelled eggs. They do not have teats. Instead, the young lick milk that oozes out of their mother's mammary glands onto a patch of skin on her belly. Two types of monotreme live in Australia — the platypus and the echidna.

Contents

Introduction 2
MONOTREMES
   Platypus 4
   Echidna 6
MARSUPIALS
   Brush-tailed phascogale 8
   Yellow-footed antechinus 8
   Fat-tailed dunnart 9
   Kowari 9
   Eastern quoll 10
   Northern quoll 11
   Spotted-tailed quoll 11
   Tasmanian devil 12
   Numbat 14
   Bilby 16
   Eastern barred bandicoot 17
   Koala 18
   Common wombat 20
   Northern hairy-nosed wombat 21
   Southern hairy-nosed wombat 21
   Eastern pygmy-possum 22
   Mountain pygmy-possum 22
   Honey possum 23
   Striped possum 23
   Sugar glider 24
   Yellow-bellied glider 25
   Common ringtail possum 26
   Common spotted cuscus 26
   Common brushtail possum 27
   Eastern grey kangaroo 28
   Red kangaroo 29
   Common wallaroo 30
   Yellow-footed rock-wallaby 30
   Red-necked wallaby 31
   Whip-tail wallaby 31
   Lumholtz's tree-kangaroo 32
   Rufous bettong 33
   Quokka 33
PLACENTAL MAMMALS
   Spectacled flying-fox 34
   Little red flying-fox 35
   Grey-headed flying-fox 35
   Diadem leafnosed-bat 36
   Hoary wattled bat 36
   Ghost bat 37
   Plains rat 38
   Spinifex hopping-mouse 38
   Greater stick-nest rat 39
   Water rat 39
   Prehensile-tailed rat 40
   Queensland pebble-mound mouse 41
   Dingo 42

Mammal conservation 44
Glossary 46
Index 47

Koalas are one of Australia's best-known mammals.

Marsupials

Young marsupials have two stages of growth. First, they grow inside the mother's belly for a short time. When they are born, they are blind, fur-less and helpless and must continue to grow inside the mother's pouch. Kangaroos, possums and wombats are all marsupials.

Placental mammals

These mammals give birth to well-developed young. They are attached to the mother by an umbilical cord, which provides nutrients to the baby while it grows inside the mother. Young placental mammals develop in a bag called a *placenta*. Humans are placental mammals and so are dingoes, bats and *rodents*.

FACTS ON THE FACT FILE RANGE

Forty-eight different species of mammal are included in this fact file. The introduction to each species helps you discover a fascinating fact about that mammal. Each profile then tells you how a species looks and behaves, what it eats, where it lives, how it *breeds* and cares for its young, if it has any *predators* or threats and, finally, its status, which means whether it is common or is in danger of becoming *extinct*. Words throughout the book that are in italics are included in the glossary on page 46 to help you increase your knowledge.

Platypus *Ornithorhyncus anatinus*

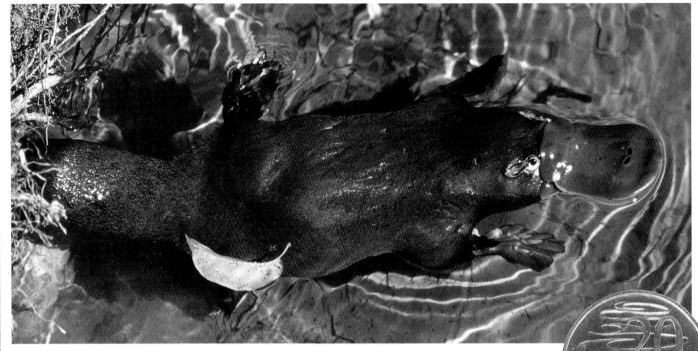

The male platypus is the only venomous mammal in Australia.

These shy, secretive monotremes only live in Australia. They have an *amphibious* lifestyle, spending about twelve hours a day searching for food in the water. The rest of the time they spend on land, either resting in camping burrows, preening or even sunbaking! The best time to see a platypus in the wild is at dusk or dawn. They bob up to the water's surface, paddle around for about ten seconds, and then dive back down to search for food.

WHAT DO THEY LOOK LIKE? The platypus is unique. Its streamlined body is covered in thick, brown, slightly oily fur that keeps it warm and dry. Its "duck-like" bill is soft, flexible and very sensitive. Its paddle-shaped tail is used as a rudder to steer the platypus when it swims. The tail also stores fat that provides energy when food cannot be found. Its webbed front feet are used to pull the platypus through the water, while the back feet are half-webbed and used for steering. Males have a hollow, *venomous* spur on each ankle, which they use to fight with other males during the breeding season.

The egg of a platypus is about 1.7 centimetres long.

SIZE: On average, males are 50 centimetres long and weigh around 1.5 kilograms. Females are smaller. They are about 43 centimetres long and weigh a kilogram.

WHAT DO THEY EAT? A platypus relies on its sensitive bill to feel for food and detect small amounts of electricity produced by its living *prey*. It sifts through the creek bed and searches under rocks, eating insects, small fishes and frogs. Platypuses store food in their cheek pouches until they come to the surface. A platypus pushes food to the back of its mouth and grinds it up between its top and bottom jaws because it does not have teeth. A platypus can eat over half its body weight in one night.

WHERE DO THEY LIVE? Platypuses live in freshwater creeks, rivers and dams in eastern Australia between Tropical North Queensland and Tasmania, across to south-eastern South Australia.

BREEDING & CARING FOR YOUNG: A female platypus digs a burrow up to 20 metres long into the side of a creek bank.

A platypus can hold its breath for about two minutes while hunting underwater.

She makes a nesting chamber at the end and lines it with soft leaves carried with her tail. She lays up to three soft, sticky eggs and curls up to *incubate* them between her tail and belly for about ten days until they hatch. A platypus doesn't have teats. Instead, young platypuses lick milk that oozes from two patches of skin on their mother's belly. The young stay in the burrow for about six weeks until they are fully furred. Their mother looks after them until they are about four months old.

PREDATORS & THREATS: Their biggest threats are freshwater pollution and *habitat* destruction to make way for human development. Predators include snakes, goannas, dingoes, spotted-tailed quolls, feral cats and foxes.

WHAT IS THEIR STATUS? They are *vulnerable*.

A platypus's bill is flexible, very sensitive and helps the platypus find its prey.

Platypuses build tunnels in creek banks just above the waterline.

The webbing on the platypus's front feet can be folded back to help it walk on land and dig burrows.

Echidna *Tachyglossus aculeatus*

An echidna's prickly spines are made of stiffened hair.

Although known simply as an echidna, this monotreme is more correctly called the short-beaked echidna. Another larger species, the long-beaked echidna, lives in Papua New Guinea. Echidnas don't make their homes in any one place; instead, they nestle under the closest pile of leaves in thick bushes, hollow logs or empty burrows. In the wild, echidnas have few predators and may live for up to 45 years.

WHAT DO THEY LOOK LIKE? Apart from its face, snout and belly, an echidna's body is covered in spiky spines of stiffened hair.

Fur in between the spines keeps its body warm. In Tasmanian echidnas, this fur may be so long that it even hides the spines. Echidnas can be light brown, reddish-brown, or even black in colour. Their feet have strong claws. Claws on the hind feet are long, curve backwards and are used for combing the fur in between the spines.

SIZE: An echidna is 30–45 centimetres long and weighs 2–7 kilograms.

WHAT DO THEY EAT? Echidnas eat ants and termites. They use their excellent sense of smell to find prey and use their strong front claws to tear open mounds and nests. An echidna's thin tongue is about 18 centimetres long and is covered in sticky saliva. It flicks its tongue in and out of a nest up to 100 times a minute, lapping up its insect prey. Echidnas don't have teeth, instead they grind their prey between hard pads at the back of the mouth. Echidnas usually feed at dawn and dusk but, on cool days, they can be seen during the middle of the day.

WHERE DO THEY LIVE? The echidna is Australia's most *widespread* mammal and is found from high snow-covered mountains to hot, dry deserts — they don't like extreme temperatures and *hibernate* to survive icy winters. In the desert, they come out at night to feed and spend the day hiding in cool, rocky crevices and caves.

Above: An echidna searches for food with its sensitive beak. *Inset above left:* Echidnas drink water but also lick dew from plants.

When swimming, an echidna uses its nose as a snorkel.

BREEDING & CARING FOR YOUNG: An echidna only breeds every 3–5 years and lives alone for most of its life. But at the beginning of the July–August breeding season, up to ten males form a "train" behind a female. For as long as six weeks, the echidnas walk, rest and feed together. When the female is ready to *mate,* she crouches down near a tree and the males fight each other until the strongest male has chased all the others away.

About two weeks after mating, the female lays an egg in a simple pouch. A fur-less young echidna, called a *puggle,* hatches about ten days later and begins to lick the rich, pink milk that seeps through its mother's skin. Once the puggle becomes too prickly for the mother's pouch, she digs a burrow and leaves it there while she goes out in search of food. She returns every few days to give her puggle a drink. At about seven months of age, the young echidna must look after itself.

An echidna's long tongue is covered in sticky saliva.

PREDATORS & THREATS: An echidna's prickly protection is enough to put off most predators, but dingoes sometimes eat the adults. The young are sometimes eaten by goannas. When threatened, an echidna curls up into a tight, spiky ball to protect its soft belly. In loose soil it can burrow straight down, leaving only its spines poking above the ground.

WHAT IS THEIR STATUS? Secure.

When threatened, an echidna can quickly burrow into the ground.

Being so low to the ground can make echidnas hard to see.

Brush-tailed phascogale *Phascogale tapoatafa*

Brush-tailed phascogales are *carnivorous* marsupials.

The brush-tailed phascogale is an active carnivore that runs up tree trunks, scampers along branches and even leaps up to 2 metres from tree to tree as it searches for food. Brush-tailed phascogales can dart forwards and backwards very easily because they can turn their back feet around 180 degrees to point behind them.

WHAT DO THEY LOOK LIKE? The most obvious feature of a brush-tailed phascogale is its furry tail. It has a grey body and creamy white belly, with big round eyes and large, fur-less ears.

SIZE: The body is about 19 centimetres long and the tail is about 20 centimetres long.

WHAT DO THEY EAT? They eat centipedes, spiders, large ants, cockroaches and other insects.

WHERE DO THEY LIVE? These marsupials are found throughout eastern, northern and south-western Australia in tall, dry forests of *eucalypts* and open woodlands that have little ground cover.

BREEDING & CARING FOR YOUNG: Males live for just one year, dying after their first mating season. Just before giving birth, a fold of skin around the female's eight teats swells to form a "part-time" pouch. She carries 3–8 fur-less young for seven weeks before leaving them in a leaf-lined nest and caring for them until they are about five months old.

WHAT IS THEIR STATUS? Secure, but vulnerable in Victoria as much of their habitat has been cleared.

Yellow-footed antechinus *Antechinus flavipes*

These *nocturnal* marsupials are comfortable living close to humans.

The yellow-footed antechinus is one of Australia's most colourful small marsupials. It is quite comfortable living with humans, and sometimes even darts around inside a house as it hunts. If it is hunting on the forest floor, it scampers along with its nose buried in the leaf litter. When it hunts in trees, rather than running along the top of the branch, it runs underneath!

WHAT DO THEY LOOK LIKE? Its head is covered in grey fur. Its feet, belly, sides and rump are ginger. A ring of white fur surrounds each eye and the tail has a black tip.

SIZE: The male's body is about 12 centimetres long and his tail is 10 centimetres long. The female's body is 10.5 centimetres long and her tail is 8.5 centimetres long.

WHAT DO THEY EAT? They hunt insects and other *invertebrates*, as well as house mice and small birds.

WHERE DO THEY LIVE? They live in north-eastern and coastal Queensland, through central New South Wales and Victoria into south-eastern South Australia and the top of South-West Western Australia.

BREEDING & CARING FOR YOUNG: All males die after mating. About a month later, the female gives birth to as many as twelve young and carries them in her pouch until they are five weeks old. The young live in the nest with their mother until the next year's breeding season.

WHAT IS THEIR STATUS? Secure.

Fat-tailed dunnart *Sminthopsis crassicaudata*

The fat-tailed dunnart stores fat in its tail to use as a food source.

WHAT DO THEY LOOK LIKE? They have large ears, bulging eyes, long whiskers and fat, carrot-shaped tails. They have soft, grey-brown fur on top and grey to white bellies.

SIZE: Males and females have a body that is 6–9 centimetres long. They have a tail that adds an extra 4–7 centimetres.

WHAT DO THEY EAT? Fat-tailed dunnarts eat invertebrates such as moths, beetles, grasshoppers and other insects.

WHERE DO THEY LIVE? These small marsupials prefer to hunt in dry woodlands, tussock grasslands, low shrublands, farms, even aridlands across southern, central and south-western Australia.

BREEDING & CARING FOR YOUNG: Fat-tailed dunnarts breed from May–June. After a twelve-day pregnancy, females usually give birth to 8–10 young. Often only about five young survive. These will leave home by ten weeks of age. Females sometimes raise two litters in a breeding season.

WHAT IS THEIR STATUS? Common.

In hot, dry habitats, fat-tailed dunnarts dodge the sun's fierce heat by hiding in cracks in the soil, under tussocks of grass or in nests beneath rocks and logs. In chilly weather, they huddle up to stay warm. Like some other carnivorous marsupials, a dunnart's thick tail works like a "lunch box" and stores fat that provides energy when food is hard to find.

Kowari *Dasyuroides byrnei*

The kowari is a nocturnal predator.

WHAT DO THEY LOOK LIKE? Kowaris are usually grey and the ends of their tails are covered in bushy, black hairs. They have large eyes and ears and long, sensitive whiskers.

SIZE: The body is about 18 centimetres long. The tail is about 12 centimetres long.

WHAT DO THEY EAT? Kowaris hunt insects and other small animals, such as mammals and geckoes. They also *scavenge* for *carrion*, or the bodies of dead animals. When there is plenty of juicy food available they do not need to drink water.

WHERE DO THEY LIVE? Small groups of kowaris are spread across south-western Queensland and north-western South Australia where there is very little ground cover.

BREEDING & CARING FOR YOUNG: Folds of skin on the mother's belly swell up to form a pouch just before she gives birth to her young. When her young become too large, they are either left in the nest or carried on their mother's back.

WHAT IS THEIR STATUS? Vulnerable.

The kowari may be small, but it is a fierce defender of its home range. A kowari threatens predators and other kowaris that come too close by hissing and making noisy, chattering sounds. Just like an angry cat, it will twitch its tail, and even lift one front foot off the ground to show it means business!

Eastern quoll *Dasyurus viverrinus*

Above: Young quolls are left in grass-lined dens while their mothers hunt.
Inset right: Up to six newborn quolls are raised in the pouch.

Eastern quolls are nocturnal hunters, but can be seen "sunbaking" during the day. Unfortunately, these beautiful animals are now only found in Tasmania. The last sighting of an eastern quoll on the mainland was in a Sydney suburb in 1963. Since then, there have been some reported sightings in Victoria and New South Wales, which leaves a glimmer of hope that one day these delightful animals might be rediscovered in these areas!

WHAT DO THEY LOOK LIKE? Eastern quolls can be either a caramel colour or black. The fur is dotted with white, but there are no spots on the tail. They have pink noses, bright eyes, large ears and needle-sharp teeth.

SIZE: The male's body is 45 centimetres long and his tail is 28 centimetres long. Males can weigh up to 1.3 kilograms. Females are slightly smaller.

WHAT DO THEY EAT? Eastern quolls hunt at night, pouncing on insects as well as bandicoots, rats, other small mammals and any birds that nest on the ground. They are quite bold scavengers and often dart into a group of feasting Tasmanian devils to steal scraps of meat. Sometimes they even eat fruit and grass.

WHERE DO THEY LIVE? Eastern quolls live throughout Tasmania from the mountains to the coast. They prefer to live where eucalypt forests meet farmland, because they feed on insects living in farm crops. During the day the eastern quoll sleeps in a den that may be under rocks, logs or in a burrow.

BREEDING & CARING FOR YOUNG: During the breeding season, the female's six teats become surrounded by skin that swells up to form a pouch. The female gives birth to up to 30 fur-less, 6-millimetre-long babies, but only the first six to attach to a teat have a chance of surviving. When they are too big for the pouch, the mother leaves them in a grass-lined den while she hunts at night. When moving from den to den, she gives her babies a piggy-back.

PREDATORS & THREATS: Foxes, feral cats and dogs are the eastern quoll's main predators. Feral cats also compete with eastern quolls for food.

WHAT IS THEIR STATUS? Secure in Tasmania but thought to be extinct on the mainland.

Northern quoll *Dasyurus hallucatus*

Northern quolls are small but savage hunters.

Northern quolls are the smallest and most aggressive of Australia's quoll species. These nocturnal predators spend more time climbing trees than any other quoll.

WHAT DO THEY LOOK LIKE? Their fur ranges from grey-brown to brown and is covered in white spots. There are no spots on the tail and, like other quolls, the belly is creamy white. Each back foot has a thumb to help the quoll climb and "ribbing" underneath for extra grip, allowing a northern quoll to run across slippery surfaces like trees and rocks.

SIZE: Males and females are both 30–31 centimetres long. The tail is about as long as the head and body combined.

WHAT DO THEY EAT? They eat *native* rats, small mammals, frogs, reptiles, grasshoppers, termites, beetles, moths, even honey and soft fruit such as figs.

WHERE DO THEY LIVE? They shelter in hollow tree trunks and termite mounds across northern Australia into Western Australia and in Queensland.

BREEDING & CARING FOR YOUNG: All male northern quolls die after the mating season finishes in late June. The female gives birth to up to eight young that suckle milk from a teat in her pouch until they are about ten weeks old. Their eyes are still closed but they already have spots on their fur.

PREDATORS & THREATS: Habitat destruction threatens their survival, and the poisonous cane toad is believed to cause many deaths. Quolls mistake cane toads for frogs.

WHAT IS THEIR STATUS? Endangered.

Spotted-tailed quoll *Dasyurus maculatus*

This is the only species of quoll that has spots on its tail.

Spotted-tailed quolls are the largest carnivorous marsupials on the Australian mainland, and they have a vicious reputation. They have large, powerful jaws with an impressive set of sharp teeth.

WHAT DO THEY LOOK LIKE? Like other quolls, the spotted-tailed quoll's chocolate brown fur is covered in white spots, but it is the only quoll to have spots on its tail. The pads on its back feet are ribbed and each foot has a "thumb" for extra grip.

SIZE: Males weigh up to 7 kilograms. They grow to 76 centimetres long, with a tail that is a 55 centimetres long. Females weigh 4 kilograms. They grow up to 45 centimetres long, with a tail that is 42 centimetres.

WHAT DO THEY EAT? They hunt other mammals, such as possums, gliders and rats, as well as birds, lizards, snakes and insects. They also scavenge carrion and hunt live prey the size of a small wallaby!

WHERE DO THEY LIVE? They make their dens along the coast and nearby mountains from South-East Queensland to Victoria and throughout Tasmania. A separate group lives in far north Queensland.

BREEDING & CARING FOR YOUNG: Usually only five young spend three weeks in the pouch before they are left in the den. Young quolls leave the den when they are eighteen weeks old.

WHAT IS THEIR STATUS? Vulnerable in Tasmania and in south-eastern Australia. Endangered in north Queensland.

Tasmanian devil *Sarcophilus harrisii*

A mother Tasmanian devil can raise up to four young from one litter.

Commonly called "Tassie devils", these mammals are the largest marsupial carnivores in the world and their screams and snarls give them a nasty reputation! Fossil records show that Tasmanian devils lived on mainland Australia until about 600 years ago. It is believed that dingoes preyed upon the smaller devils and caused their disappearance. Luckily there are no dingoes in Tasmania.

WHAT DO THEY LOOK LIKE? These stocky marsupials have large heads with very strong jaws. They have short, thick tails that store fat in case food is hard to find. Their black and white colour helps them *camouflage* as they hunt at night. They have bright red ears and long whiskers.

SIZE: The height of a large male can be up to 30 centimetres and they can weigh up to 12 kilograms. Females are a little smaller and weigh around 10 kilograms.

WHAT DO THEY EAT? Their Latin name, *Sarcophilus*, means "flesh-loving" and Tasmanian devils are true carnivores. They like to scavenge dead animal *carcasses*, or carrion, eating almost half their own body weight during a single meal.

Where there is not a lot of carrion, Tasmanian devils are active predators, eating anything they can hunt. Young devils pounce on tadpoles, moths, frogs, birds and lizards. Adults eat these as well, but they also hunt pademelons, wallabies and wombats. The Tasmanian devil can devour an entire carcass (even the bones) using its strong, powerful jaws. After a big meal, Tasmanian devils can go without food for 2–3 days. They also play an important role in decreasing the risk of flies and disease because they clean up the carcasses.

Above: Male Tasmanian devils often fight over females during the mating season. *Left:* A Tasmanian devil's teeth and jaws are very powerful.

FEEDING BEHAVIOUR: Up to 30 devils gather around a carcass where they snap, snarl, scream and growl at each other. Tasmanian devils need to work out a "pecking order", which means that the loudest, scariest, hungriest devil gets to eat the carcass first. Although their squabbles look very aggressive, Tasmanian devils rarely hurt each other during fights for food. Often females with young devils to feed are the bossiest.

WHERE DO THEY LIVE? They live in all Tasmanian habitats, even outer-city suburbs and farms. They are most common in dry eucalypt forests and coastal woodlands. Devils sleep in a simple den, often a hollow log, and may share a few dens in their home range with other devils.

BREEDING & CARING FOR YOUNG: A female gives birth to around 30 young but she only has four teats in her backwards-facing pouch. The fur-less babies are born in April and spend four months inside the pouch, drinking milk and developing. When fully furred, they are left in the den and begin to practise hunting and survival skills. Young devils leave the den in December.

PREDATORS & THREATS: Because there are no dingoes in Tasmania, fully grown devils do not have any natural predators. However, young devils must be careful both day and night. Wedge-tailed eagles can snatch them during the day, while masked owls and spotted-tailed quolls hunt them at night. Even other hungry devils are dangerous. Another serious threat is Devil Facial Tumour Disease, a cancer that kills adult devils by causing large, lumpy tumours to grow on their faces and necks. This disease affects the devil's teeth and stops it feeding. Scientists are still trying to work out how to prevent the disease.

WHAT IS THEIR STATUS?
Common, but some scientists recommend changing their status to vulnerable.

Tasmanian devils rely on their sense of hearing when hunting prey. They also have an excellent sense of smell.

Inset above: Young devils playfighting.

Young devils often have tug-of-wars over food.

Numbat *Myrmecobius fasciatus*

A numbat has 4–11 white stripes across its back.

The numbat is *diurnal* — it sleeps at night in a nest and wakes up to search for its food during the day. It was once the world's most endangered mammal, but numbers are slowly beginning to increase with some help from humans protecting the numbat's habitat.

Inset above: Numbats do not have pouches.

Numbats have large bushy tails and striped, colourful fur. The striped pattern breaks up the body's outline and makes them hard to see.

WHAT DO THEY LOOK LIKE? Numbats have white stripes across their black rumps. They have pointy noses and fluffy tails. The rest of the fur is reddish-brown. A dark stripe runs from their eyes to the base of their ears.

SIZE: Males sometimes grow larger than females, but usually they are about the same size. The head and body is about 24 centimetres long and the tail is about 17 centimetres long.

WHAT DO THEY EAT? Numbats only eat termites, but sometimes lick up ants at the same time.

Numbats are active diggers that spend the day burrowing in search of termites.

The striped colour pattern helps numbats camouflage themselves and confuse predators.

FEEDING BEHAVIOUR: The numbat is a busy little daytime digger. It trots along, nose to the ground, with its bristly tail curled above its back. It stops when it detects termite activity under sticks and branches or just below the surface, stopping to dig and uncover termite tunnels. The termites stick to the numbat's long, flickering tongue — up to 20,000 may be eaten in one day!

WHERE DO THEY LIVE? Numbats are now only found in two small regions in South-West Western Australia, where eucalypt forests and woodlands provide plenty of hollow logs for nesting, sleeping and escaping predators. These habitats also have a good supply of termites. Numbats are burrowers that line their nests in flowers, soft leaves and torn-up bark.

BREEDING & CARING FOR YOUNG: Numbats are marsupials but they do not have a pouch. A female gives birth to furless, 1-centimetre-long young that attach themselves to one of her four teats. Dangling from her belly, the young numbats are dragged along until they are about six months old and 5 centimetres long. When they are too big to carry, the mother leaves them alone in the burrow while she searches for termites during the day. At night, she sleeps in the burrow and feeds her young milk. When they are about ten months old, the young numbats begin life on their own.

PREDATORS & THREATS: Numbats are eaten by carpet pythons, dingoes and diurnal birds of prey such as the little eagle. The clearing of their habitat for farming is also a major threat to the numbat.

WHAT IS THEIR STATUS? Vulnerable.

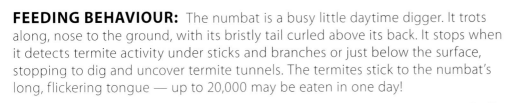
Numbats need lots of hollow logs for nesting and sleeping.

Bilby *Macrotis lagotis*

Bilbies are also known as rabbit-eared bandicoots.

Bilbies are the only type of bandicoot to live underground where it is cool and where they can escape the heat of the day. Using its strong paws, the bilby digs a three-metre-long winding burrow two metres below the surface. Bilbies have long back feet like a kangaroo but, rather than hopping, they run along the ground on all fours. Their long, rabbit-like ears help keep them cool as blood passes close to the skin's surface and lets extra heat escape.

WHAT DO THEY LOOK LIKE? The fur is soft, silky and blue-grey. The belly is white. The tail is black with long white hairs at the end and a fur-less tip. The bilby has a pointed, pink nose and long rabbit-like ears.

SIZE: Males weigh about 2.5 kilograms and are about 55 centimetres long. Females weigh 1.25 kilograms and are 39 centimetres long. The tail is about 28 centimetres long on both males and females.

WHAT DO THEY EAT? Bilbies do not have good eyesight but have excellent smell and hearing. They prey on grasshoppers, beetles, termites, fungi, spiders, seeds and fruit, and get most of their water from their juicy food. They dig down into the soil for fleshy plant stems called bulbs, and also lick up seeds with their long, thin tongues.

WHERE DO THEY LIVE? Bilbies once lived throughout most of Australia's aridlands, but are now only found in the Great Sandy, Tanami and Gibson Deserts where the soils are rocky or sandy and covered in *spinifex* or acacia shrub. A group also survives in the dry grasslands of south-western Queensland.

BREEDING & CARING FOR YOUNG: The bilby has a backwards-facing pouch. One or two tiny young spend about ten weeks in the pouch, drinking milk and developing, before spending a further two weeks in the burrow. The mother goes out at night in search of food, but comes back a few times each night to feed her young milk. A bilby begins to breed from six months of age.

WHAT IS THEIR STATUS?
Endangered.

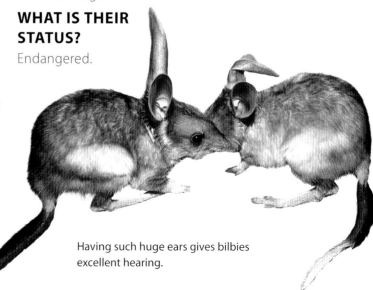

Having such huge ears gives bilbies excellent hearing.

Eastern barred bandicoot *Perameles gunnii*

At night, eastern barred bandicoots pop out of their grassy nests and go hunting for food. These little marsupials dig shallow, cone-shaped holes in the soil and poke their noses in to smell for food. They pounce on any prey they find, crunching it up with their needle-sharp teeth.

WHAT DO THEY LOOK LIKE?
The eastern barred bandicoot has wiry, yellow-brown fur with three or four striking pale stripes across the rump and a short, white, pointy tail. This bandicoot has a long, pointed snout and, like other bandicoots, has an excellent sense of smell and hearing.

SIZE: Both males and females grow to 35 centimetres long. Males and females weigh between 0.5 kilograms and close to 1.5 kilograms.

WHAT DO THEY EAT? Earthworms are one of their favourite meals, but beetles, other insects and grubs are also eaten, as are berries and fungi. If their food is juicy enough, they do not need to drink water.

WHERE DO THEY LIVE? Eastern barred bandicoots live in northern and eastern Tasmania, and in one area close to Hamilton in Victoria. They live in grassy woodlands and grasslands close to shelter and farm paddocks. In Victoria, some even live in backyard gardens.

Bandicoots begin to venture out of their grassy nests just after dusk.

BREEDING & CARING FOR YOUNG: Adults live alone and only meet up with other bandicoots to breed. A female has a backwards-facing pouch and, on average, carries three young that drink her milk as they develop. These bandicoots begin to breed as soon as they are about fourteen weeks old, and don't live longer than two years.

WHAT IS THEIR STATUS? Endangered on mainland Australia. Vulnerable in Tasmania.

Zoos and sanctuaries breed bandicoots in *captivity*. They release them back into natural environments that are safe from feral predators.

Koala *Phascolarctos cinereus*

Koalas are very fussy about the type of gum leaves they eat.

The koala is one of Australia's most familiar animals. With its thick shaggy fur, big black nose and round, fluffy ears it might look like a bear, but koalas are in fact marsupials. Koalas spend about 20 hours a day sleeping. The rest of the time they move through the canopy eating eucalyptus leaves. They are agile climbers, with strong claws on their hands and feet that are perfect for this job.

WHAT DO THEY LOOK LIKE? Koalas are grey and woolly with creamy-white bellies. However, koalas that live in the cold southern regions are larger with darker, fluffier fur. Koalas hands have two thumbs and three fingers, each with a long, sharp claw to help grip smooth branches. Their big toes are clawless and work like thumbs. The next two toes are joined together to form one toe with two claws. This is used for scratching and combing the fur to make it waterproof.

Inset right: This young koala will stay in the pouch until it is covered in fur and fully developed. *Right:* A young koala too big for its pouch rides on its mother's back.

SIZE: In the south, male koalas weigh up to 15 kilograms and grow to around 80 centimetres long. Large females weigh about 11 kilograms and measure just over 70 centimetres long. The smaller, northern males weigh up to 9 kilograms and grow to 74 centimetres long. Northern females weigh just over 7 kilograms and are about 72 centimetres long.

WHAT DO THEY EAT? Koalas eat up to a kilogram of gum leaves every day. Even though there are more than 600 different species of gum tree growing in Australia, koalas are very fussy and only eat the leaves of a few.

FEEDING BEHAVIOUR: Gum leaves are tough and provide the koala with little energy. However, a koala has a 2-metre-long "caecum", like a human's appendix, which is full of millions of tiny bacteria that help the koala digest the leaves and give it energy. There is enough moisture in gum leaves and koalas do not need to drink. Gum leaves are poisonous to most animals, but the koala's liver breaks these *toxins* down so the koala does not get sick.

WHERE DO THEY LIVE? Koalas live wherever there is enough food and living space for them. Although their ancient relatives lived in rainforests, koalas now live in tall eucalypt forests, inland woodlands and even coastal islands, from Kangaroo Island off South Australia up along the east coast and inland to central north Queensland. Koalas live wherever their favourite types of gum leaves are found.

BREEDING & CARING FOR YOUNG: A newborn koala joey is blind and fur-less. It has strong front paws to drag itself into its mother's backwards-facing pouch where it spends the next five months drinking milk from one of her two teats. When it becomes covered in fur, it will take its first look outside the pouch. To get the young koala's belly ready for a diet of leaves, the mother produces special, soft droppings called "pap", which the baby licks. It then begins to nibble gum leaves and learns which ones to eat. Koalas do not make a nest, so when the joey is too big for the pouch, it rides on its mother's back. When the joey is about one year old, the mother gives birth to another joey.

PREDATORS & THREATS: Pythons, goannas, dingoes and large predatory birds such as eagles are natural predators of young koalas. When adult koalas travel along the ground between trees they may also be attacked by dingoes. However, the biggest threat to koalas is the clearing of their habitat or breaking it into small areas so there is a lot of ground between one group of trees and the next. Koalas can then be attacked by dogs or hit by cars as they move between trees. When koalas become stressed, they become sick and infected with an eye disease that causes blindness.

WHAT IS THEIR STATUS? Secure.

A joey faces many predators and threats. Until a joey is old enough to fend for itself, it always stays close to its mother for protection.

Koalas' hands have two thumbs for extra grip while climbing.

Common wombat *Vombatus ursinus*

A wombat uses its strong front paws and claws to dig its burrow.

Wombats have one of the best-developed brains of any marsupial. However, they are stubborn and if something is in their way they simply bulldoze it down. This can sometimes include branches and even fences! At night they keep their noses to the ground in search of food, but at the slightest noise or threat of danger, they burst off in a flurry at speeds of 40 kilometres an hour!

WHAT DO THEY LOOK LIKE? Common wombats have large heads and stocky bodies with short powerful legs and very strong claws. Their short fur is wiry and may be brown, grey or black. The nose is large and fur-less. A wombat's rump is covered in thick, tough skin that predators find hard to bite or scratch through.

SIZE: Both males and females can grow to over a metre long and weigh almost 40 kilograms.

WHAT DO THEY EAT? Wombats have poor eyesight and rely on their good sense of smell to find their food, which includes the leaves of shrubs, tree roots, soft mosses and native grasses. They even sniff out food hidden under snow.

WHERE DO THEY LIVE? Common wombats live in mountainous woodlands, coastal *heathlands*, open eucalypt forests and grasslands from southern Queensland through New South Wales to Victoria, South Australia and Tasmania.

Wombats often roll on the ground to cover themselves in dust.

They usually dig burrows into the sides of hills. A burrow may be up to 5 metres deep with a network of branching tunnels up to 30 metres long. The burrow has several entrances and may be shared by another wombat. Common wombats are nocturnal but sometimes venture out during the day when it is cool or cloudy.

BREEDING & CARING FOR YOUNG: A female's pouch faces backwards so when she digs her young isn't covered in sticks, rocks and dirt. A baby wombat stays in the pouch drinking milk until it is about ten months old. It then spends another 8–10 months by its mother's side until it is ready to start life on its own.

PREDATORS & THREATS: Tasmanian devils, dingoes and wedge-tailed eagles hunt wombats. To escape, the common wombat races into a hollow log or burrow and blocks the entrance with its hard rump. If the predator gets its head in between the wombat and the ceiling of the log, the wombat pushes its rump up and crushes the predator's skull!

WHAT IS THEIR STATUS? Secure.

During winter, wombats may come out in daytime to rest in the sun.

Once a young wombat is fifteen months old, it no longer needs its mother's milk.

Northern hairy-nosed wombat *Lasiorhinus krefftii*

The Epping Forest population of wombats was discovered in 1937.

Wombats are found nowhere else in the world and of the three different species, the northern hairy-nosed is the largest! Sadly, these sturdy little diggers are one of Australia's most endangered mammals. There are just over 100 left surviving in one small patch of forest surrounded by a dingo-proof fence — none are kept in captivity.

WHAT DO THEY LOOK LIKE? The northern hairy-nosed wombat has soft, silky grey-brown fur with a large, square-shaped, hairy nose. This wombat also has long, pointed ears.

SIZE: They are slightly larger than common wombats, growing to more than a metre in length and weighing an enormous 40 kilograms!

WHAT DO THEY EAT? These wombats nibble on the leaves of native grasses. They only spend 2–6 hours feeding above ground. They spend the rest of the day in their burrows trying to save energy.

WHERE DO THEY LIVE? Northern hairy-nosed wombats live in a tiny 300-hectare patch of Epping Forest National Park in Central Queensland.

BREEDING & CARING FOR YOUNG: After the baby is born it makes its way into the backwards-facing pouch and drinks milk for about six months before starting to nibble blades of grass. By nine months old it is out of the pouch, but stays close to its mother.

WHAT IS THEIR STATUS? Endangered.

Southern hairy-nosed wombat *Lasiorhinus latifrons*

A southern hairy-nosed wombat at the entrance to its burrow.

Like all wombats, southern hairy-nosed wombats are *herbivores* — they eat plants but also eat a lot of gravel and dirt! Amazingly, a wombat's teeth never wear down to stumps because they don't have roots like human teeth. Instead, they keep growing throughout the wombat's life. In fact, a wombat needs to keep grinding its teeth away so they do not become too long.

WHAT DO THEY LOOK LIKE? The soft, silky fur of the southern hairy-nosed wombat may be grey to reddish-brown. The nose is square-shaped and covered in fine white hairs.

SIZE: Adults weigh 19–32 kilograms and grow to 93 centimetres long.

WHAT DO THEY EAT? The split, top lip of the southern hairy-nosed wombat helps it chew grasses right down to the stem. There is hardly any water where these wombats live, so they get all the moisture they need from their food.

WHERE DO THEY LIVE? These wombats live in grasslands and open eucalypt and acacia woodlands in areas of South Australia and across parts of the Nullarbor Plain just into Western Australia.

BREEDING & CARING FOR YOUNG: A southern hairy-nosed wombat is ready to breed at the age of three. One baby is born and stays in the pouch for up to nine months. By one year of age, the baby no longer feeds on milk — it eats grass just like an adult.

WHAT IS THEIR STATUS? Secure in their small range.

Eastern pygmy-possum *Cercartetus nanus*

Eastern pygmy-possums enjoy the nectar of native flowers.

Pygmy-possums are tiny, which is why they got the name "pygmy". They live in the trees and are excellent climbers. During winter they curl up, tuck their big ears in, and go to sleep. Called "torpor", this is a way for these possums to survive through a time when there is not a lot of food available.

WHAT DO THEY LOOK LIKE? The thick, fluffy fur is grey-brown and the belly is white or grey. Some even have a red, rusty-coloured head. They have large, bright eyes with big, mouse-like ears. The long tail is scaly and has only a thin covering of hairs. Their two back paws have a big toe that works like a thumb to grip twigs as they run. If there is plenty of food around, the base of the tail begins to store fat.

SIZE: Adults grow 7–11 centimetres long. The tail is almost the same length.

WHAT DO THEY EAT? These little possums are *omnivores* with a large range of tastes. They dip their brush-tipped tongues into native flowers to lick up nectar and pollen. They have sharp teeth and eat juicy insects such as moths and grasshoppers. When trees aren't flowering, they feed on soft fruit.

WHERE DO THEY LIVE? These pygmy-possums live from South-East Queensland down the coast into south-east South Australia and throughout Tasmania.

BREEDING & CARING FOR YOUNG: Although females have six teats in the pouch, only four young are raised in the litter. The young stay in the pouch for about a month before they start sleeping in the nest.

WHAT IS THEIR STATUS? Secure in most places, but vulnerable in New South Wales.

Mountain pygmy-possum *Burramys parvus*

Mountain pygmy-possums eat bogong moths.

Once thought to be extinct, these pygmy-possums were rediscovered in 1966 around a mountain ski resort. They are the largest of the pygmy-possums and the only Australian mammal to live high in the mountains where it is freezing cold and snowing. They hibernate during the cold and sometimes even tuck or bury seeds in secret holes. They eat these later when food is hard to find.

WHAT DO THEY LOOK LIKE? This possum has soft, thick grey fur with a creamy-coloured belly. When breeding, the belly and sides turn a bright yellow-orange. The long, scaly tail is *prehensile* and used for gripping. Each eye is surrounded by a dark ring of fur.

SIZE: The mountain pygmy-possum's body is around 14 centimetres long. Its tail is around 12 centimetres long.

WHAT DO THEY EAT? They eat bogong moths, beetles, caterpillars, spiders and centipedes, as well as the seeds and fruit of high country shrubs.

WHERE DO THEY LIVE? They live in the high country of Mt Bogong, Mt Hotham and Kosciuszko National Park, at least 1400 metres above sea level. The habitats in these areas are made up of heath and low shrubs.

BREEDING & CARING FOR YOUNG: Females give birth to a litter of four young, which spend a month in the pouch and a month in the nest before they are ready to leave. The young then start breeding the following spring.

WHAT IS THEIR STATUS? Endangered.

Honey possum *Tarsipes rostratus*

Honey possums only eat nectar from flowers.

Honey possums have many interesting features. They have the least teeth of any marsupial (less than eleven), females give birth to the tiniest young of any mammal (weighing only five-thousandths of a gram), but males have the longest *sperm* of all mammals (about one-third of a millimetre). They hold twigs in a monkey-grip and have fingernails rather than claws.

WHAT DO THEY LOOK LIKE? The light brown or grey fur has three darker stripes running down the back. The nose is long and pointy. It has a long, prehensile tail and, as in all possums, the second and third toes are joined, with claws for combing the fur.

SIZE: Females are larger than males. Large females grow to about 9 centimetres long, with a tail that is 11 centimetres long. Females weigh a tiny 12 grams.

WHAT DO THEY EAT? Honey possums are "nectarivores", which means that they only eat pollen and nectar from flowers. They poke their pointy noses down into a flower and stick their bristly, brush-tipped tongues inside to collect their sweet food.

WHERE DO THEY LIVE? They are only found throughout the sandy heathlands, open woodlands and forests of southern Western Australia where there are plenty of grass-trees.

BREEDING & CARING FOR YOUNG: Some females have at least two litters of up to four young every year, but usually less. They suckle young in the pouch for two months before leaving them in the nest.

WHAT IS THEIR STATUS? Secure.

Striped possum *Dactylopsila trivirgata*

These possums look and smell like skunks.

Rather than racing along branches at night, the striped possum crashes its way through the leaves at high speed, snorting and leaping from tree to tree. It uses two razor-sharp teeth to tear open bark and find prey. It also uses its extra-long fourth finger to poke into holes and stab hard-to-get prey.

WHAT DO THEY LOOK LIKE? This possum has bright, striking black and white stripes covering its dainty body. These stripes form a white "Y" shape on the face. The long tail often has a white tip and the fourth finger on both hands is much longer than the others. It has a bright pink nose and a strange, musty smell.

SIZE: Large adults grow up to 27 centimetres long. The tail is 34 centimetres long.

WHAT DO THEY EAT? Striped possums eat all kinds of insects and grubs. They also eat small reptiles and nibble on leaves. They have also been known to lap up the honey from native bee hives.

WHERE DO THEY LIVE? Striped possums live in rainforests and eucalypt woodlands along the east coast of Cape York Peninsula in Tropical North Queensland.

BREEDING & CARING FOR YOUNG: A female striped possum has two teats in her pouch and raises as many as two young. However, not a lot is known about this possum's breeding.

WHAT IS THEIR STATUS: Secure.

Sugar glider *Petaurus breviceps*

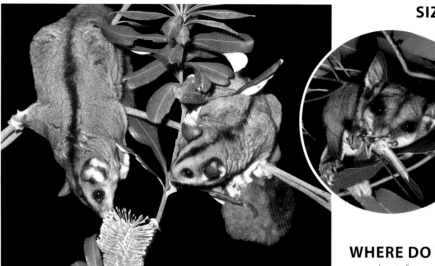

Above and inset right: Sugar gliders lick nectar and eat insects.

Sugar gliders can move from tree to tree by gliding 50 metres through the air. A thin piece of skin, called a *patagium*, stretches all the way from the little finger to the ankle. When the sugar glider stretches out its arms and legs, the patagium makes it look like a "kite". Its long, fluffy tail helps to steer the glider through the air.

WHAT DO THEY LOOK LIKE? The sugar glider has incredibly soft, grey fur, a pink nose, bulging, bright eyes and large ears with a black tip. The belly is creamy-coloured and a dark stripe runs from the middle of the face, along the back and down to the long, fluffy tail. The tail sometimes has a white tip, but it is not prehensile, so the glider cannot hang by its tail from a branch.

SIZE: The tail and body of the sugar glider are about the same length, both 16–21 centimetres long. Males are only slightly bigger and heavier than females.

WHAT DO THEY EAT? A sugar glider uses its sharp front teeth to crunch up insects and seeds, and scratch the bark of trees, which ooze sugary, sweet sap. They lick nectar and pollen from flowers with their soft pink tongues.

WHERE DO THEY LIVE? Sugar gliders live in eucalypt forests, woodlands and rainforests across northern Australia, down the coast to south-eastern South Australia and throughout Tasmania.

BREEDING & CARING FOR YOUNG: Sugar gliders huddle together in a large family group in a leaf-lined nest inside a hollow tree branch. A female gives birth to two young. She carries them in her pouch for just over two months. She feeds them in the nest until they are four months old. They then begin to head out with her to learn how to find food. The young leave the nest at 7–10 months of age.

PREDATORS & THREATS: Many young gliders don't survive their first year of life. Adults are hunted by pythons, quolls, owls, goannas and kookaburras. *Introduced* predators such as foxes and feral cats also kill them. Trees with hollows are very important to their survival.

WHAT IS THEIR STATUS? Secure.

Above: Sugar gliders live together in large family groups. *Inset above:* Hollow trees are very important to the survival of the sugar glider.

Yellow-bellied glider *Petaurus australis*

Yellow-bellied gliders spend almost the entire night busily searching for food, running upside-down along branches and sometimes travelling up to 2 kilometres from their daytime dens. They use the patagium to help them glide through the canopy, easily covering distances up to 120 metres from one tree to the next. Gliding is the best way to quickly escape lurking predators on the ground below, and to save energy.

WHAT DO THEY LOOK LIKE? These large gliders are covered in light or dark grey fur. An adult's belly may be yellow, cream or white, but a young glider's belly is always paler. The tail is fluffy and much longer than the tail of any other glider. They have large, furless ears, black paws, and a patagium that stretches from the wrist to the ankle.

SIZE: Males are heavier than females, but both may grow up to 30 centimetres long. The tail on both males and females grows to 48 centimetres long!

Yellow-bellied gliders bite chunks out of tree trunks to lick the sap.

WHAT DO THEY EAT? Sap and nectar make up most of their diet. To make a tree ooze sap, a yellow-bellied glider bites a 5-centimetre-long, V-shaped chunk out of a trunk or branch with its sharp teeth. These gliders also eat the pollen of eucalypt flowers and any insects and spiders living in the trees or hiding under the bark.

WHERE DO THEY LIVE? This large glider lives in forests of tall, mature eucalypts between south-eastern South Australia and north Queensland. Their habitat must have hollow trees for shelter and nesting, and must contain plenty of food.

BREEDING & CARING FOR YOUNG: A female's pouch is divided into two separate areas, each with its own teat; however, only one young is born at a time. The young glider lives in the pouch for about 100 days before being left in the den at night while its mother searches for food. The young glider continues to drink its mother's milk for another two months before starting life on its own.

PREDATORS & THREATS: The main threats to this glider are the cutting down of tall, old trees that have hollows big enough for yellow-bellied gliders to live in. Foxes and feral cats prey on these gliders. Large owls and carpet pythons also eat yellow-bellied gliders.

WHAT IS THEIR STATUS? Vulnerable.

Yellow-bellied gliders can soar for distances up to 120 metres.

Common ringtail possum *Psuedocheirus peregrinus*

These possums are common visitors to suburban backyards.

This shy, nocturnal possum has a thin tail, which is prehensile and is used as a fifth arm for grabbing onto tree branches. The tail is also used to curl around and carry shredded bark and grass back to the possum's nest.

WHAT DO THEY LOOK LIKE? The tail has a white tip and a fur-less strip of skin underneath, which works like a non-slip pad. The fur is mostly rusty, reddish-brown but can be grey to almost black. The belly is white. They have short, rounded ears with a white patch behind them. Their front paws are "syndactylous", which means they can move their second finger over to their first. This gives them extra grip while climbing, and is just like having two thumbs.

SIZE: The body and tail are about the same length. Both are 30–35 centimetres long.

WHAT DO THEY EAT? Common ringtail possums eat different types of leaves — sometimes they even eat eucalypt leaves. They also eat flowers and fruit of many native trees and shrubs, as well as those of introduced plants, like roses.

WHERE DO THEY LIVE? They can be found right across Cape York Peninsula, down along the east coast and into south-east South Australia.

BREEDING & CARING FOR YOUNG: These are the only possums where the male helps to look after the young. A female has four teats, but gives birth to two young. They leave the pouch when they are four months old. To help the female, the male guards them in the nest while the mother is out searching for food.

WHAT IS THEIR STATUS? Secure.

Common spotted cuscus *Spilocuscus maculatus*

The cuscus also has a curly, prehensile tail to grip onto branches.

The common spotted cuscus is a nocturnal feeder but, during winter, it enjoys lying in the sun during the day. This cuscus has very woolly fur for a creature that lives in the hot, steamy tropics. To cool its body down, the common spotted cuscus pants like a dog and wets its fur-less feet and face with saliva — this works in the same way as sweating.

WHAT DO THEY LOOK LIKE? They have thick, fuzzy fur with tiny ears and bulging eyes. Only males have spots on their grey fur, and females may have a white-yellow patch on the rump. The last two-thirds of the tail is completely fur-less and covered in little bumps for extra grip.

SIZE: Adults grow to 58 centimetres long. The tail is about 44 centimetres. They weigh almost 5 kilograms.

WHAT DO THEY EAT? The common spotted cuscus is a herbivore that eats rainforest fruit, leaves and flowers.

WHERE DO THEY LIVE? The common spotted cuscus lives across the tip of Cape York Peninsula in its preferred habitat of rainforests. They are also found in fresh and saltwater mangroves, paperbark forests and eucalypt woodlands up to 500 metres away from the nearest rainforest.

BREEDING & CARING FOR YOUNG: Females have as many as three young, but it seems that only one young survives to be raised outside the pouch.

WHAT IS THEIR STATUS? Secure, but they are scattered throughout their range.

Common brushtail possum *Trichosurus vulpecula*

The common brushtail possum is Australia's most familiar possum. They are bright-eyed, bushy-tailed and bold. Human development hasn't decreased the number of these marsupials. Instead, they have become regular visitors to camping grounds, backyards and back verandahs. Sometimes they even make their homes in the ceilings of houses and noisily race across roofs at night, chasing each other. Mothers are often seen piggy-backing their babies around as they search for food.

WHAT DO THEY LOOK LIKE? The fur is usually silver-grey, but in cooler areas it may be almost black and shaggy. In more tropical habitats, the fur is orange-red in colour and shorter. The golden colour of some common brushtail possums is caused by a *mutation* in the *genes*. The tail is prehensile and, with the help of the fur-less strip underneath, supports the possum's body as it hangs from a branch. Their ears are round and about 6 centimetres long.

SIZE: Adults grow up to 55 centimetres long. The bushy tail is 40 centimetres long. Males weigh up to 4.5 kilograms. Females weigh up to 3.5 kilograms.

Common brushtails use their excellent sense of smell to find food

WHAT DO THEY EAT? Brushtails are omnivores that mainly eat leaves, fruit and flowers. When they can, they eat insects and sometimes rob nests of their eggs and even baby birds.

WHERE DO THEY LIVE? Their main habitats are open eucalypt forests, woodlands, rainforests and areas around human homes. Common brushtail possums were introduced into New Zealand and have become feral because they have no natural predators.

BREEDING & CARING FOR YOUNG: A female begins to breed at one year of age. A blind, fur-less joey spends up to five months in the pouch, attached to one of two teats as it drinks milk and grows fur. It then climbs out and is carried on its mother's back for another month or two, still suckling, but also beginning to eat what its mother eats.

PREDATORS & THREATS: Dingoes, large goannas and pythons are natural predators, but foxes, domestic dogs and cats also chase and kill these native mammals. Hollows in trees are very important and, in backyards, nest boxes make good homes. They also encourage these possums to move out of people's ceilings.

WHAT IS THEIR STATUS? Secure.

Above: A golden-coloured common brushtail possum. *Inset right:* The golden colour of these possums is caused by a mutation of their genes.

Eastern grey kangaroo *Macropus giganteus*

Above: A kangaroo's tail helps it keep balance while it is hopping. *Inset right:* The teat swells up inside the joey's mouth so that it has a constant supply of milk while in the pouch.

The first European settlers were very surprised when they first saw large, grey animals hopping along on two legs rather than running on four. Hopping at high speed on spring-loaded legs is a lot easier than running, and a much better way to save energy. Eastern grey kangaroos hold the marsupial speed record at 64 kilometres an hour!

WHAT DO THEY LOOK LIKE? Males are strong and muscular with light to dark grey fur. Females have white furry chests and pouch areas. The ends of the feet, paws and tail may be black. Eastern grey kangaroos have long back feet and powerful back legs that work together as they hop or move. The tail is long, stiff and very muscular.

SIZE: These are Australia's second-largest kangaroo. Males stand about 1.6 metres tall and weigh up to 70 kilograms. Females are about 1.2 metres tall and weigh 35 kilograms.

WHAT DO THEY EAT? Kangaroos are known as *crepuscular* feeders — most of their feeding is done in the early morning and late afternoon when it is much cooler. Large groups of eastern greys, called mobs, graze on different grasses and ground plants with large leaves. They spend the rest of the day lying down in the shade under trees and shrubs.

WHERE DO THEY LIVE? Eastern greys need well-watered grasses where rainfall is more than 25 centimetres per year. They are found on coastal and open inland plains, along the Great Dividing Range and in the north-east corner of Tasmania.

BREEDING & CARING FOR YOUNG: A newborn joey is tiny. It is fur-less and blind, and does not have proper ears or a tail. It has only tiny stumps for back legs, but has well-developed, strong arms. It pulls itself up into the pouch and attaches itself to one of its mother's four teats. Straight away the teat swells in the joey's mouth so that it cannot fall off and the joey can continue drinking milk. After about nine months, the joey is covered in fur and ready to leave the pouch for short periods. At eleven months it still continues to come back to the pouch to drink from the same teat. It does this until it is about eighteen months old. By this time, there is already another joey in the pouch that is about eight months old.

PREDATORS & THREATS: Dingoes and domestic dogs may kill and injure young kangaroos. Large birds of prey also hunt them. In the early days of farming, kangaroo numbers increased as more water became available for stock and they began to graze on crops. Hunters were paid money to hunt kangaroos to stop them becoming a pest, and sold the skins and meat.

WHAT IS THEIR STATUS? These kangaroos are secure on the mainland, but vulnerable in Tasmania.

Above: Young male kangaroos often "box" and playfight with one another as they grow.

Left: Once it is covered in fur, a joey begins to poke its head out of its mother's pouch.

Red kangaroo *Macropus rufus*

Red kangaroos can also be grey in colour.

The red kangaroo is not only Australia's largest native marsupial, but it is also the largest marsupial in the world — standing as tall as a man. These magnificent marsupials are the only kangaroo to live in the arid outback where it is very hot and drought is always a concern.

WHAT DO THEY LOOK LIKE? Male red kangaroos have rusty-red fur and are called "boomers", while the females are grey-blue and sometimes called "blue flyers". However, some large males are grey-blue while some females may be red. They all have white fur on the belly. You can easily tell red kangaroos apart from eastern grey kangaroos by the broader nose and mouth. This area, called the "muzzle", has black and white patches on each side and a wide, white stripe that runs up to the ears.

SIZE: A large male can stand up to 2 metres tall and weigh up to 85 kilograms. Although a female stands about 1.3 metres in height, she only weighs up to 35 kilograms.

WHAT DO THEY EAT? Red kangaroos are crepuscular. They begin feeding just before dawn, rest during the day, and begin to feed again in the late afternoon and into the evening. When it is available, they nibble on short, sweet, green grasses and native herbs and even the leaves of some shrubs. Mobs are usually small, but if plenty of food and water is available after its rains, hundreds of red kangaroos may group together.

WHERE DO THEY LIVE? Red kangaroos live across most of central and western Australia in areas where it is very dry. Their preferred habitats are deserts, open woodlands and grasslands. To escape the heat of the day, red kangaroos find shade to rest and save energy. To help get rid of extra body heat, they also pant like dogs and lick the insides of their forearms with saliva, which then cools in the breeze.

BREEDING & CARING FOR YOUNG: Red kangaroos breed when conditions are good and there is enough food and water for the females and their young to survive.

If conditions are good, a female may have three young in different stages of growth at the one time. She can have a *fertilised* egg in her belly just waiting to develop into a tiny, fur-less joey, another joey in her pouch suckling on one teat, and another joey, too big for the pouch but still coming back for milk, drinking from another teat.

PREDATORS & THREATS: Dingoes are natural predators of the red kangaroo.

WHAT IS THEIR STATUS? Secure.

Inset right: A joey keeps going back inside its mother's pouch until it is about eight months old.

Common wallaroo *Macropus robustus*

Wallaroos don't form large mobs like kangaroos.

A wallaroo is a marsupial that is between the size of a kangaroo and a wallaby. The common wallaroo is the most widespread *macropod*. A macropod is a big-footed marsupial that hops. Kangaroos, wallaroos and wallabies are all macropods. Unlike their larger kangaroo relatives, wallaroos like to live alone rather than in mobs.

WHAT DO THEY LOOK LIKE? The common wallaroo is a stocky marsupial. Males are often dark grey while females and young males are blue-grey. The common wallaroo has a white belly and a large, black, fur-less nose.

SIZE: Females stand about 1.2 metres tall and weigh up to 25 kilograms. Males are stockier, weighing close to 47 kilograms, and standing up to 1.6 metres tall.

WHAT DO THEY EAT? Their main food is grass, but they also eat plants that aren't very nutritious, like spinifex, as well as shrubs and other ground plants.

WHERE DO THEY LIVE? Common wallaroos live around rocky slopes, which are great places to rest in the cool shade. The common wallaroo's range covers most of Australia. They survive in deserts, areas of eucalypt forest, woodlands and grasslands.

BREEDING & CARING FOR YOUNG: Common wallaroos give birth to one joey every year, but do not breed during droughts. A joey spends about nine months developing in its mother's pouch.

WHAT IS THEIR STATUS? Secure.

Yellow-footed rock-wallaby *Petrogale xanthopus*

Yellow-footed rock-wallabies are hunted by wedge-tailed eagles.

This is the largest of the rock-wallabies and also the prettiest and most colourful of all the macropods. Because of their beautiful fur, early settlers killed thousands for the fur-trade, greatly decreasing their numbers. Now foxes are a major predator, while feral goats eat the same food and shelter in the same caves and rocky places as the wallabies that need to escape the 40-degree summer sun.

WHAT DO THEY LOOK LIKE? The fur on the back is grey-brown and the belly is white. The ears, arms, legs and tail are orange-yellow. The tail is long with brown bands across it. Like all rock-wallabies, their back feet are short and padded. The skin is rough and stops this wallaby slipping on smooth surfaces.

SIZE: Adults stand about 60 centimetres tall. The tail is 70 centimetres long. These wallabies weigh up to 11 kilograms.

WHAT DO THEY EAT? Yellow-footed rock-wallabies eat grasses and leaves, including the branches of shrubs during times of drought.

WHERE DO THEY LIVE? These wallabies live in semi-arid areas where there are many rocky cliffs to shelter from the hot sun. The Flinders Ranges in South Australia is where the largest number live. Scattered colonies also live in western Queensland and in a small area of western New South Wales.

BREEDING & CARING FOR YOUNG: Females reach breeding age at 1–2 years of age.

WHAT IS THEIR STATUS? Vulnerable in South Australia. Endangered in New South Wales.

Red-necked wallaby *Macropus rufogriseus*

Red-necked wallabies are known as Bennett's wallabies in Tasmania.

Red-necked wallabies don't live in mobs like most other kangaroos and wallabies. Instead, they prefer to live alone; however, they sometimes gather in groups of more than 30 when they are feeding at night. Once a joey leaves the pouch — rather than staying by its mother's side — it tucks itself away in thick undergrowth while she feeds.

WHAT DO THEY LOOK LIKE? Most of the fur is a grey-brown colour, with a rusty-red tinge around the neck and sometimes on the rump. The tips of the paws and feet are black. The fur of the Tasmanian red-necked wallaby is thicker, darker and shaggier to help it cope with cold, snowy weather.

SIZE: Males stand up to a metre tall and weigh 20 kilograms. Females stand about 80 centimetres tall and weigh 14 kilograms.

WHAT DO THEY EAT? Grass forms the main part of the red-necked wallaby's diet, but it also nibbles other ground plants and leaves when it needs to.

WHERE DO THEY LIVE? There are two populations — one lives on the mainland's east coast between south-east South Australia and Central Queensland. The other group lives in Tasmania and on islands in the Bass Strait.

BREEDING & CARING FOR YOUNG: Females on the mainland breed at any time of year, as long as there is plenty of food. The young leave the pouch when they are nine months old.

WHAT IS THEIR STATUS? Secure.

Whip-tail wallaby *Macropus parryi*

Whip-tail wallabies often sit with their tails between their legs.

The whip-tail wallaby has dainty features and white cheek-stripes. Also known as the "pretty-face" wallaby, it usually lives in small groups of up to ten, but mobs of more than 100 gather when there is lots of food available. If a wallaby senses danger, it thumps its long foot on the ground to let the others know. They then all take off in different directions to confuse predators.

WHAT DO THEY LOOK LIKE? The face has chocolate-brown fur covering the muzzle. The chest and belly are white. The rest of its fur is grey to brown.

SIZE: Males stand about 1.2 metres tall and weigh 26 kilograms. Females are only 85 centimetres tall and weigh 15 kilograms.

WHAT DO THEY EAT? They eat grass, some ground plants and ferns. They don't need to drink, except through droughts, because their food is juicy enough and they also lick dew. They feed during the early morning, late afternoon and into the night.

WHERE DO THEY LIVE? These wallabies live in open eucalypt forests where the ground is covered in plenty of grass. They live along eastern Australia from north Queensland to northern New South Wales.

BREEDING & CARING FOR YOUNG: Females begin breeding when they are two years old, while males begin mating at the age of three. The young joey is carried in the pouch until it is close to nine months old, but comes back to suckle for another six months.

WHAT IS THEIR STATUS? Secure.

Lumholtz's tree-kangaroo *Dendrolagus lumholtzi*

Lumholtz's tree-kangaroos live in the trees, but sometimes hop on the ground. These tree-climbers can move their back legs one at a time and can even walk backwards, unlike ground-living kangaroos. As the Lumholtz's tree-kangaroo sits on a branch, its long tail hangs straight down and helps it balance, but it cannot be used to grip onto branches like a possum's tail.

WHAT DO THEY LOOK LIKE? The Lumholtz's tree-kangaroo has much shorter legs and stronger arms than its ground-living relatives. The feet are rectangular in shape and are very well padded underneath. The skin under each foot is bumpy and works like a non-slip pad as the kangaroo moves along smooth branches. The face, hands, feet and tail are dark brown to black. The body fur is grey to black-brown with a pale rump.

SIZE: This is the smallest of Australia's two species of tree-kangaroo. Adults grow to around 52–65 centimetres long. The tail is 65–73 centimetres long. Males weigh around 8.5 kilograms. Females weigh up to 7 kilograms.

A tree-kangaroo's arms and legs are short and strong.

WHAT DO THEY EAT? Tree-kangaroos are mainly "folivores", which means they eat leaves. They also eat rainforest fruit and sometimes come to the ground, hopping into nearby farm paddocks and eating corn.

WHERE DO THEY LIVE? They live in tropical rainforests of north Queensland in areas higher than 800 metres above sea level. They live alone and instead of building nests, sleep during the day crouched on branches high in the forest canopy.

BREEDING & CARING FOR YOUNG: A female only gives birth to one young, but has four teats. The joey attaches itself to the most swollen teat and stays in the pouch for about eight months. Once out of the pouch, it stays with its mother until it is two years old.

PREDATORS & THREATS: They don't have many predators, but large snakes like the amethystine python may hunt tree-kangaroos. The main threat is the destruction of rainforest homes by clearing.

WHAT IS THEIR STATUS? Secure.

Above: The tree-kangaroo's long tail helps it to balance as it sits on a branch. *Right:* Tree-kangaroos mainly eat the leaves of rainforest trees, but sometimes eat fruit.

Rufous bettong *Aepyprymnus rufescens*

The rufous bettong has strong front claws for digging up fungi.

The rufous bettong is not only the largest, but also the most widespread, of the group of small hopping marsupials known as rat-kangaroos. It is nocturnal and spends the day sleeping in one of the five or six globe-shaped nests it weaves from grass. Each nest is built about 100 metres from the next.

WHAT DO THEY LOOK LIKE? The silvery-grey fur has a reddish tinge. The rufous bettong has a white belly, pink, pointy ears and a wide, hairy nose. It has a white tail, long thin feet and long claws.

SIZE: Both males and females stand about 40 centimetres tall. Males weigh about 3 kilograms while females may weigh 3.5 kilograms.

WHAT DO THEY EAT? About half an hour after the sun drops below the horizon, this bettong begins its search for fungi, grass, flowers, roots, seeds and even the bones of other animals.

WHERE DO THEY LIVE? They are most commonly found between the coastal regions of north Queensland and northern New South Wales and around the Murray River. They live in open eucalypt forests and woodlands that have grassy (not thick and bushy) floor coverings.

BREEDING & CARING FOR YOUNG: One young is born, spending two months attached to a teat. It continues to share a nest with its mother and stays by her side until it is about six months old.

WHAT IS THEIR STATUS? Secure.

Quokka *Setonix brachyurus*

People travel from all over the world to see quokkas on Rottnest Island.

The quokka has become a major tourist attraction of Rottnest Island in Western Australia. When a Dutch explorer first discovered the island in 1696, he was amazed at its creatures that looked like a "kind of rat". He described the island as a "rat nest", which is where the name "Rottnest" comes from. The island gives quokkas a safe home away from foxes that live on the mainland.

WHAT DO THEY LOOK LIKE? The long, reddish-brown fur is wiry and flecked with grey. The quokka has a short tail, small, rounded ears and a dark, fur-less nose. The body is round and solid.

SIZE: Standing upright, adults are about 30 centimetres tall. Males can weigh just over 4 kilograms. Females weigh about 3.5 kilograms.

WHAT DO THEY EAT? Quokkas eat grass and leaves. By the end of summer on Rottnest Island, many quokkas die because there is not enough nutritious food or drinking water close by.

WHERE DO THEY LIVE? The quokka only survives in large numbers on Rottnest Island off Perth and Bald Island off the southern Western Australian coast. On the mainland, it lives in tiny populations around swamps outside Perth and in South-West jarrah forests.

BREEDING & CARING FOR YOUNG: One young is born and cared for in its mother's pouch. The joey spends about six months in the pouch, then another three outside drinking milk from one of the teats.

WHAT IS THEIR STATUS? Vulnerable.

Spectacled flying-fox *Pteropus conspicillatus*

Bats are the only mammals that can fly.

Spectacled flying-foxes are easy to recognise by the yellow fur that forms circles around their eyes. In Australia, they are the only one of the twelve different species of large bat (or "megabat") to rely on the rainforest for their survival. However, the rainforest also needs the flying-foxes for its survival. As they move through the forest, the flying-foxes play an important role in spreading the seeds of different trees.

WHAT DO THEY LOOK LIKE? These flying-foxes have dark brown to black fur, with yellow-orange eye rings and neck scruff. The scruff can also be silver. The legs have fur down to the knees. The arms have four long fingers and a thumb covered in thin, stretchy skin that forms wings, and joins to the ankles. The thumb works like a hook for climbing along branches.

SIZE: The head and body of an adult is 22–24 centimetres long.

WHAT DO THEY EAT? They lick nectar and pollen from eucalypts and other flowers, but feed mainly on the fruit of certain rainforest trees. Because they fly at night and find their food by smell and sight, they usually suck on light-coloured fruit that is easier to find in the dark.

FEEDING BEHAVIOUR: When feeding, the spectacled flying-fox always hangs upside down.

WHERE DO THEY LIVE? By day these bats roost in large colonies, called "camps", in north Queensland rainforests, mangroves and paperbark swamps. They tend to live no further than 6 kilometres from the closest rainforest.

BREEDING & CARING FOR YOUNG: A female gives birth to one fully furred baby with its eyes already open. The baby clings to its mother's fur with its sharp claws and latches onto one of the teats under each wing-pit with its tiny milk teeth. Even though young flying-foxes can fly at about two months of age, they are left in the colony and cared for until they are about five months old.

PREDATORS & THREATS: Flying-foxes are eaten by large pythons, owls and even crocodiles. The main threat to their survival is from clearing of their rainforest habitat to make room for human development and farms. Flying-foxes love fruit, and the farms that replace their natural habitat grow tasty tropical fruit like bananas, lychees, and mangoes. Many bats are hunted, electrocuted or poisoned because they eat these fruit crops.

WHAT IS THEIR STATUS? Vulnerable.

Inset above right: A flying-fox colony. *Above:* When resting, flying-foxes hang upside down and wrap their wings around their bodies like a blanket.

Little red flying-fox *Pteropus scapulatus*

Little red flying-foxes roost together in large colonies called camps.

Little red flying-foxes roam the countryside, following the path of flowering eucalypt trees. Not only does this behaviour make them Australia's most widespread fruit bat, they are also found further inland than any other species. These bats eat a lot of food from the forest but they also *pollinate* flowers and help new trees to grow.

WHAT DO THEY LOOK LIKE? The body fur is red-brown but may have pale yellow on the neck and shoulders. The legs are fur-less and the skin on the red-brown wings is so thin it is almost see-through.

SIZE: These bats grow to around 24 centimetres long and weigh 300–600 grams.

WHAT DO THEY EAT? These bats mainly feed on the nectar and pollen of flowering eucalypt trees. But fruit, leaves, sap, and even bark and insects are also eaten. These bats also visit orchards to feast on fruit crops.

WHERE DO THEY LIVE? More than a million little red flying-foxes may roost together in the one camp. They are usually close to water in mangroves, woodlands and paperbark forests along coastal and inland regions of northern and eastern Australia and on Kangaroo Island off South Australia's coast.

BREEDING & CARING FOR YOUNG: A female gives birth to one young, which is cleaned, suckled and taught how to find the best food until it is at least four months old.

WHAT IS THEIR STATUS? Secure.

Grey-headed flying-fox *Pteropus poliocephalus*

 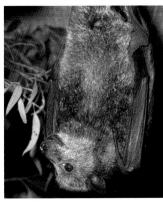

The grey-headed flying-fox is found further south than any other flying-fox.

It is a common sight at dusk to see a dark cloud of flapping wings fill the sky as grey-headed flying-foxes search for food. Their noisy chatter soon gives way to the sound of their flapping wingbeats as they travel at speeds of up to 35 kilometres an hour. Flying-foxes have excellent eyesight, hearing and smell, and can pick up the scent of flowering trees and fruit many kilometres away!

WHAT DO THEY LOOK LIKE? The head is grey and the body is covered in shaggy, dark brown or grey fur. The neck fur is rusty-red. The legs have fur right down to the ankles. Black skin covers the wings.

SIZE: The *wingspan* is around a metre. The body is 23–29 centimetres long. These bats weigh up to a kilogram.

WHAT DO THEY EAT? They eat the nectar and pollen of eucalypt flowers and also ripe fruit.

WHERE DO THEY LIVE? These flying-foxes are found in rainforests, coastal heaths, swamplands or eucalypt forests from Bundaberg in Queensland to Melbourne in Victoria and inland across the Great Dividing Range.

BREEDING & CARING FOR YOUNG: Around October, six months after mating, the female gives birth to a fully furred baby that clings to her belly and drinks from a teat in each of her wing-pits. After about three weeks it stays in the camp at night while its mother flies out to feed. The baby practises flying and food-finding at about three months of age.

WHAT IS THEIR STATUS? Vulnerable.

Diadem leafnosed-bat *Hipposideros diadema*

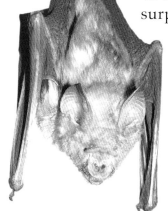

Diadem leaf-nosed bats are *ambush* predators with large, sharp teeth. They silently hang from branches above paths where insects fly. Here they wait, ready to capture their prey by surprise. Like other small bats (or *microbats*), they use *echolocation* to find prey. They make sounds that cause invisible waves in the air. These sounds hit an insect and "echo" back to the bat, letting it know how big the insect is, how fast it is flying and how far away it is.

As they hang, these bats twist from side to side to locate prey.

WHAT DO THEY LOOK LIKE? The fur is grey, yellow-brown or bright orange. The flaps of skin around the nostrils form a *noseleaf*, which helps to funnel out high-pitched sounds and help the bat find prey. The large, pointy ears have small ridges of skin that look like stripes.

SIZE: Males and females grow 7.5–8.5 centimetres long. The tail is 3–4 centimetres long.

WHAT DO THEY EAT? They eat beetles, bugs, ants and moths.

WHERE DO THEY LIVE? These bats live in caves in the tropical rainforests and eucalypt forests of coastal north Queensland and Cape York Peninsula.

BREEDING & CARING FOR YOUNG: Females have four teats and give birth to one baby. A teat on each side of the chest feeds the baby milk, but there is another two teats between the legs. These do not release milk. They are for the baby to bite and hold onto when being carried in flight. Once the baby becomes too heavy to carry, it is left in the cave and protected by other young bats.

WHAT IS THEIR STATUS? Vulnerable.

Hoary wattled bat *Chalinolobus nigrogriseus*

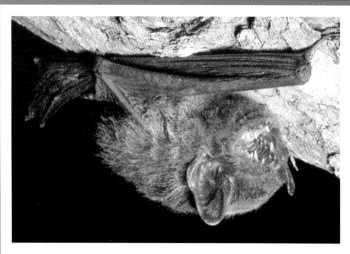

Microbats have sharp teeth to crunch the hard bodies of insect prey.

As dusk falls, hoary wattled bats leave their daytime roosts, before any other microbats set out, to begin their night of hunting. Their strange name refers to the way they look. The white tips on the fur give them a frosted, or "hoary" look. Between the ears and the mouth are pieces of dangly skin called "wattles".

WHAT DO THEY LOOK LIKE? The fur is dark, smoky-grey. Each hair has a white tip that is longer on some hairs than others. A thin piece of skin, like that on the wings, runs from one ankle to the tip of the long, thin tail and back across to the other ankle.

SIZE: These tiny bats grow to 4.5–5.5 centimetres long. The tail is just over 4 centimetres long.

WHAT DO THEY EAT? These bats eat insects such as moths, grasshoppers and crickets. They also hunt other invertebrates such as spiders. They snatch much of their prey in mid-air but if they detect non-flying prey below, they swoop down, land on the ground and use their wings to crawl after the prey and catch it.

WHERE DO THEY LIVE? These bats live right across northern Australia down to the mid-east coast of New South Wales. They stay away from arid habitats, and roost in cracks between rocks and tree hollows in moist forests and open woodlands.

BREEDING & CARING FOR YOUNG: It is thought that one young is born in spring.

WHAT IS THEIR STATUS? Vulnerable in New South Wales. Secure elsewhere.

Ghost bat *Macroderma gigas*

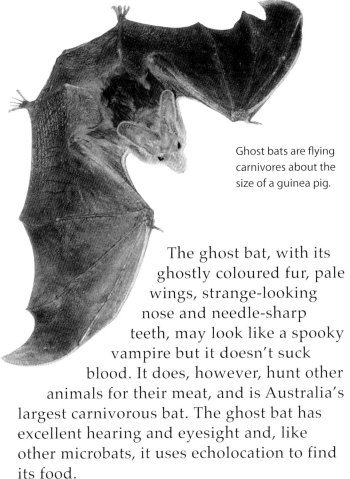

Ghost bats are flying carnivores about the size of a guinea pig.

The ghost bat, with its ghostly coloured fur, pale wings, strange-looking nose and needle-sharp teeth, may look like a spooky vampire but it doesn't suck blood. It does, however, hunt other animals for their meat, and is Australia's largest carnivorous bat. The ghost bat has excellent hearing and eyesight and, like other microbats, it uses echolocation to find its food.

WHAT DO THEY LOOK LIKE? The body is covered in soft, light to dark grey fur, but the belly is a paler, sometimes white, colour. Ghost bats don't have tails and the wing skin stretches back between the legs. The long ears are joined together in the centre. They have large eyes and a long noseleaf.

SIZE: Adults grow 10–13 centimetres long and only weigh around 165 grams. The wingspan is about a metre wide.

WHAT DO THEY EAT? Ghost bats eat lizards, frogs, sleeping birds, large insects and small mammals including hopping mice, dunnarts and even other bats. When hunting, they often perch about 2 metres above the ground and listen for rustling in the grass below. They swoop out and use their excellent eyesight and echolocation to pinpoint the animal before dropping down and biting it behind its head or neck. The animal is then carried back to the bat's perch and devoured.

WHERE DO THEY LIVE? Ghost bats only live in Australia. They can be found in a very broad range of habitats, from lush north Queensland rainforests, up through the forests and woodlands of Cape York Peninsula and around the dry, arid regions of Western Australia. Ghost bats roost in cracks between rocks, caves and old mine shafts in colonies of just a few to more than 400 bats!

BREEDING & CARING FOR YOUNG: About six months after mating, a female ghost bat gives birth to one young, usually around October. As summer heats up, the females move their young to the warmest caves. Young ghost bats can fly at about seven weeks of age. The young bats go out with their mothers to learn hunting skills until they are able to look after themselves.

PREDATORS & THREATS: Snakes, including pythons (such as the spotted python), hunt warm-blooded animals like the ghost bat. Owls may also hunt and catch them. Ghost bats are very sensitive to human disturbances to their roosting sites. Many of these sites have been cleared to make way for mines or farms. Feral cats and foxes also compete with the ghost bat for food.

WHAT IS THEIR STATUS? Vulnerable.

Inset right: The ghost bat uses its noseleaf to help find prey.
Far right: Ghost bats roost in caves during the day.

Plains rat *Pseudomys australis*

A young plains rat leaves the nest before it is one month old.

The group known as rodents make up more than half of all the mammals on the planet. In Australia, they make up about one quarter of all mammals, and are found in just about every habitat on land. Plains rats have found a way to cope with life in the desert. Unlike humans, plains rats don't have sweat pores! They get all the moisture they need from their food and don't need to drink water.

WHAT DO THEY LOOK LIKE? The silvery-grey fur is very soft. Like all native rodents, they keep themselves very clean and don't carry harmful diseases. They have large ears and big eyes.

SIZE: Females and males grow to 10–14 centimetres long. The tail is 8–10 centimetres long.

WHAT DO THEY EAT? Plains rats eat seeds, some leaves, stems and even insects.

WHERE DO THEY LIVE? As many as 20 rats share a shallow, underground maze of connected tunnels. Once widespread, they now only live in the Lake Eyre basin close to the centre of Australia and in a small area of the Nullarbor Plain just across the South Australian border into Western Australia where the habitat is made up of dry shrublands and stony deserts.

BREEDING & CARING FOR YOUNG: These native rats only breed after heavy rains allow a good supply of food to grow. Three or four young are raised on their mother's milk and grow quickly — they are ready to leave the nest before they are even one month old!

WHAT IS THEIR STATUS? Vulnerable.

Spinifex hopping-mouse *Notomys alexis*

The spinifex hopping-mouse begins its search for food at dusk.

By day, the spinifex hopping-mouse huddles in warm, moist, underground burrows. It only comes to the surface once the sun has dropped below the horizon. While searching for food, the spinifex hopping-mouse moves around on all fours. At the slightest hint of danger it will hop away at high speed, springing over obstacles more than six times its own body height!

WHAT DO THEY LOOK LIKE? The big eyes, large, rounded ears and long tail sets this rodent apart from all others. The spinifex hopping-mouse is covered in light brown fur and has a grey-white belly. The back feet are long and the legs are powerful.

SIZE: The body is 10 centimetres long. The tail is about 14 centimetres long.

WHAT DO THEY EAT? Spinifex hopping-mice are omnivores and eat invertebrates such as insects, spiders and centipedes, or the seeds of shrubs and grasses, fresh leaf shoots and plant roots. They don't need to drink water as long as their food is juicy.

WHERE DO THEY LIVE? They live throughout the aridlands of central and Western Australia. Their habitats range from sandy deserts covered in spinifex tufts to dry woodlands and grasslands.

BREEDING & CARING FOR YOUNG: Up to six young are raised in a litter.

WHAT IS THEIR STATUS? Secure.

A hopping-mouse holds its food in its front paws as it nibbles.

Greater stick-nest rat *Leporillus conditor*

These rodents were once common but are now endangered.

The greater stick-nest rat only has two front teeth on its top jaw and two on the bottom. These teeth don't stop growing and must be continually worn down. To do this, the rat chomps its way through thick branches and builds a mound of sticks for its nest. The nest can be a metre high and 1.5 metres wide. As many as 20 rats share this nest.

WHAT DO THEY LOOK LIKE? Their soft, fluffy fur is yellowish-brown to grey. The belly is a creamy colour. The eyes and ears are large and round and the back feet are white on top.

SIZE: Adults grow up to 26 centimetres long. The tail is about 18 centimetres long.

WHAT DO THEY EAT? These rats are strict herbivores and only eat the fruit and leaves of native plants.

WHERE DO THEY LIVE? The only place they are now naturally found is in open woodlands and low shrubland on Franklin Island off South Australia.

BREEDING & CARING FOR YOUNG: These rats mate in autumn and, after about six weeks, the female gives birth to as many as four young.

PREDATORS & THREATS: These rats were once common on the mainland before the spread of sheep farming and introduced rabbits. These rats have now become extinct in these areas. Natural hunters like the dingo and predatory birds eat these rats. Foxes and feral cats also prey on them.

WHAT IS THEIR STATUS? Endangered.

Water rat *Hydromys chrysogaster*

A water rat about to devour a freshwater crayfish.

Water rats are active hunters both during the day and at night. They live an amphibious lifestyle, spending part of their day in the water and part of it on land. To help them stay dry, they have waterproof fur, but cannot spend too long in icy waters. They need to spend time sitting in the sun and shivering to warm themselves. They were once hunted for their pretty, soft fur.

WHAT DO THEY LOOK LIKE? The waterproof fur can range from dark, steely grey on top to almost black. The fur underneath can be white, or even bright orange. The long tail has a white tip. The back feet are wide and half-webbed for better paddling.

SIZE: Males grow to 31 centimetres long. The tail is 28 centimetres long and they weigh about 750 grams. Females are smaller.

WHAT DO THEY EAT? They are mainly carnivorous predators that eat fishes, frogs, insects, lizards, birds, spiders and hard-shelled yabbies or crayfish. During winter, when prey is hard to find, they also eat plants. They wade through shallow water or dive down to 2 metres to catch prey, taking it back to a favourite eating site to devour safely.

WHERE DO THEY LIVE? They live in slow-moving, freshwater creeks, rivers, swamps and dams where they nest in hollow logs or in tunnels dug into riverbanks.

BREEDING & CARING FOR YOUNG: A female is able to give birth to up to five litters of three or four young every year.

WHAT IS THEIR STATUS? Secure.

Prehensile-tailed rat *Pogonomys mollipilosus*

Above: Prehensile-tailed rats are excellent climbers and find their food in trees. *Inset left:* These rodents stay very clean by washing and grooming their fur.

The prehensile-tailed rat is thought to be Australia's most beautiful rat. Its long tail is used like an extra hand to wrap around and cling to branches as the rat climbs. The tail twists around the branch and grips on tightly as the rat stretches out to grab the next branch. If the rat slips, its tail stops it from falling — it supports its whole body weight!

WHAT DO THEY LOOK LIKE? The fur is silky-soft and grey on top. The belly is pure white. They have caramel-coloured markings on the face and a thin, black ring around each eye. The tail is long and prehensile.

SIZE: Males grow to almost 17 centimetres long. The tail is 21 centimetres long. Females grow to about 15 centimetres long. The tail is 19 centimetres long.

WHAT DO THEY EAT? Prehensile-tailed rats eat the leaves and fruit of certain rainforest trees, as well as those of introduced plants. They also feast on the fruit of pandanus, the leaves of yellow passionfruit vines and even green bananas.

FEEDING BEHAVIOUR: These rodents are nocturnal and only ever come out at night to feed. They feed in trees and bushes above the ground. When feeding, they sometimes take fruit off a branch to carry in the mouth as they dart away. When they stop, they hold the fruit in their front paws while they eat. They usually feed alone, but sometimes small groups of about three gather to feed on the leaves of yellow passionfruit vines.

WHERE DO THEY LIVE? Prehensile-tailed rats live in Cape York Peninsula and between Cooktown and Cairns — areas that are covered in rainforest.

The rat spends the day asleep in a burrow in the forest floor where it digs out a nesting chamber — layering it with soft green leaves for comfort. While the burrow only seems to have one entrance, there is always a secret escape tunnel with a hidden entrance tucked under the leaf litter. They live underground but they feed in the canopy, coming to ground every so often.

BREEDING & CARING FOR YOUNG: Very little is known about their breeding. Scientists know that although females have six teats, they only raise litters of two or three young.

PREDATORS & THREATS: Predators like pythons, northern quolls and lesser sooty owls hunt prehensile-tailed rats.

WHAT IS THEIR STATUS? Secure.

Queensland pebble-mound mouse *Pseudomys patrius*

There are five different types of pebble-mound mouse living in Australia and each of them lifts stones, sometimes as heavy as the mouse itself, piling them up to build an enormous mound around the burrow. It is the females that do most of the heavy-lifting work — the males spend their energy racing off and mating with as many females as they can.

WHAT DO THEY LOOK LIKE? The Queensland pebble-mound mouse is about half the size of a house mouse. It has orange-brown fur with black flecks and a white nose and belly. The feet are pink. The tail and whiskers are long and thin. The ears are large and rounded.

SIZE: Adult males and females grow to about 7 centimetres long. The tail is slightly longer — just over 7.5 centimetres. Males and females weigh a tiny 12 grams!

WHAT DO THEY EAT? These nocturnal mice are omnivorous, but their main diet is made up of seeds, grass and other plants. They also catch and eat small insects.

WHERE DO THEY LIVE? Because of their need for pebbles, they live along the open rocky ridges of the Great Dividing Range in north-east Queensland in habitats of low woodland and grassland. They tunnel underground and live in burrows protected by enormous, fortress-like rock walls.

MOUND-BUILDING BEHAVIOUR: These little mice pick up pebbles and carry them in the mouth to make a pile surrounding the burrow entrance. Up to a kilogram of pebbles can be shifted in one night! As they position each rock, they hammer it into place with their front paws. After many nights of heavy lifting the result is a volcano-shaped mountain of rocks — 40 centimetres wide at its base, 15 centimetres high, and having a total weight of 30–40 kilograms!

Giant mounds of rocks protect the entrance to the mouse's burrow.

BREEDING & CARING FOR YOUNG: Not a lot is known about their breeding in the wild. Some pebble-mound mice living in captivity are known to have had up to seven litters, giving birth to around 22 young in a year. Females care for their young and stay very close to the mound, while males move away and travel over a kilometre a night in search of more females. If they are lucky, males live for about 10 months. Females live for just over a year.

PREDATORS & THREATS: During their nightly travels, males are open to attack by various predators including feral cats, dogs, owls, snakes and even marsupial mice!

WHAT IS THEIR STATUS? They are *rare*.

A litter of baby pebble-mound mice.

A pebble-mound mouse uses its front paws to "brush" its fur.

Dingo *Canis lupus dingo*

There is something wild and spine-chilling about hearing the calls of a pack of howling dingoes. They are Australia's only native dog and the country's largest carnivorous mammal. Unlike domestic dogs, dingoes don't bark. Instead, they use their howls and short, sharp "yaps" to communicate with other pack members. In many areas they are widespread and common; however, pure-bred dingoes are becoming very hard to find.

WHAT DO THEY LOOK LIKE? The most familiar colour of a dingo is ginger. But they may also be black, white or tan-coloured, depending on the habitat in which they live. The paws, belly and chest are usually white and they have a white tip on the end of the fluffy tail. They are naturally thin dogs and some dingoes have black markings on the muzzle.

SIZE: Dingoes grow to about 1.2 metres long, stand just over 60 centimetres tall at the shoulder and can weigh as much as 24 kilograms. The bushy tail is almost 40 centimetres long. Males may be slightly larger than females.

WHAT DO THEY EAT? Dingoes are most active at dawn and dusk, but also hunt at night and during the day. They hunt rabbits, kangaroos, lizards, wallabies, wombats, possums, magpie geese, rodents, insects, feral pigs and goats. Dingoes also eat carrion and, if nothing else is around, will even eat fruit!

These two dingo pups are being hand-raised in a wildlife park.

FEEDING BEHAVIOUR: Dingoes are top predators that change their style of hunting to suit their prey. They work alone when hunting small prey like rabbits, but work in packs when hunting larger prey. This allows them to kill animals as large as a buffalo!

Dingoes hunt in packs to capture large prey like kangaroos.

Dingo pups wait at the entrance of their underground den.

WHERE DO THEY LIVE? Dingoes are found throughout most habitats on Australia's mainland, including deserts, alpine forests, coastal heaths and rainforests — as long as there is water and plenty of good den sites. They are found on some offshore islands, but don't live in Tasmania.

BREEDING & CARING FOR YOUNG: Dingoes often live in a family pack with as many as twelve members. A male and female breeding pair head the family while the others, usually pups from previous years, help to rear the newborns. Every year dingoes give birth to one litter of up to 10 pups. Pups are born in a den that may be underground, in a hollow log, inside a cave or at the base of thick bushes. The pups suckle milk from their mother for three or four months before they begin hunting.

PREDATORS & THREATS: Although dingoes are hunted in some farming areas because they prey on young calves and lambs, numbers are still plentiful. The greatest threat to the dingo's survival is cross-breeding with domestic and feral dogs. It is often difficult to tell the difference between a pure-bred and cross-bred dingo, and it is thought that at least 80 percent of the dingoes in eastern Australia are no longer pure-bred.

WHAT IS THEIR STATUS? Secure.

Howling is the way dingoes communicate with each other.

Fraser Island is home to many pure-bred dingoes.

MAMMAL CONSERVATION

When the first white settlers arrived in Australia, they began to knock down animal homes and clear forested areas. Shortly after, they introduced animals that were never supposed to live on the continent, and the native wildlife did not have ways of defending themselves against their invaders. Since that time, nineteen native Australian mammals have been wiped out, never to be seen again, while a further ten species are now no longer found on the mainland. In this short amount of time, more mammal species have disappeared from Australia than any other continent on Earth. Of those still surviving, around 20 percent remain in danger of disappearing.

Feral cats are nocturnal hunters that eat small mammals and birds.

WHY THE FERAL FRENZY?

The terrible impact that introduced animals have had on Australian mammals and their homes was not realised by early European settlers. Twelve rabbits released for hunting took only 50 years to reach plague proportions and cover most of Australia. Some animals, such as the cat, were even introduced to try to control rabbit numbers and, in the meantime, became a pest by breeding, and hunting native animals! Native mammals have also found it hard to compete against the fast-breeding ferals that eat the same food they do. Fortunately, we now realise the full extent of the feral problem. Hopefully, we will not only rectify this problem, but stop it from ever happening again.

This brushtail possum is safe in a Tasmanian backyard.

MAMMAL HOMES CLOSE TO HOME

The survival of native mammals depends on national parks, conservation reserves, and large areas of natural habitat on privately owned land free from feral animals. Many old trees have been chopped down to make way for human development but, in the process, the hollow homes of mammals have fallen with them. A great way of bringing mammals back into your own backyard is to give them a hollow home in the form of a nest box. Work out what lives in your area. Google the Australian web to find out the dimensions and start building a nest box today. A great place to start is www.birdsaustralia.com.au/infosheets/nestbox.html

Feral rabbits compete for habitat and food with native animals.

A HELPING HAND

It is easy for all of us to help Australia's native mammals survive. By remembering a few simple tips and lending a helping hand, not only will we help native mammals survive, we will bring ourselves much closer to them!

- Become a junior wildlife ranger and learn as much as you can about native wildlife.
- Join a local wildlife volunteer group and help monitor mammal populations in your area.
- Plant native trees and shrubs in your backyard to provide flowers, leaves and shelter for mammals.
- Protect native mammals from your cat by keeping your purring friend inside with you at night.

Zoos and wildlife sanctuaries breed native animals.

Scientific research helps us to understand wildlife.

MAMMAL "MOTHERS"

There are many people who become temporary parents to sick, injured or orphaned mammals. There are wildlife carer groups in every State. Contact one near you to find out how you can become a mammal's "mother".

Never feed wild animals. They must learn to find food for themselves.

SHOULD I FEED WILD ANIMALS?

Getting close to native mammals in their natural habitat is an exciting experience. However, trying to feed wild animals can become dangerous for both animals and people. Some animals, like the dingo, may look skinny and in need of food, but are just naturally very thin. Feeding wild animals encourages them to depend on humans. They soon become very bold and sometimes even aggressive towards humans. They may also lose their natural hunting instincts. Native animals are wild and would not be living in an area unless there was enough food for them to eat.

Wildlife carers help sick or injured mammals.

GLOSSARY

Ambush A sudden and surprise attack.

Amphibious Living both on land and in water.

Breed To produce young that can also have their own babies.

Camouflage Colours that help an animal blend into its background.

Captivity When an animal is kept in a zoo or other enclosure.

Carcass The dead body of an animal.

Carnivore An animal that eats meat and other animals.

Carrion A dead, rotting animal.

Crepuscular An animal that feeds in the early morning and late afternoon (when it is much cooler).

Diurnal Active during daylight hours.

Echolocation The use of high-pitched sounds and echoes to find objects and prey.

Eucalypt A gum tree.

Extinct Having no living member of a species.

Fertile Able to produce young.

Gene Physical characteristics that are passed on from parents to their young.

Habitat The place in nature where a certain kind of animal lives and breeds or where a plant grows.

Heathland An area of open land with scattered shrubs and small trees.

Herbivore An animal that eats plants.

Hibernate To spend the winter resting or asleep in one place.

Incubate To keep eggs warm until they hatch.

Introduced Having been brought from another country or environment.

Invertebrate An animal without a backbone.

Macropods The group of herbivorous Australian marsupials with strong hind legs (includes kangaroos and wallabies).

Mammary glands Milk-producing glands in all mammals.

Mate When animals mate, the male transfers special cells (called "sperm") to the female's eggs, which causes young animals to develop. "Mate" also means a partner.

Microbat A bat that uses echolocation to find its prey.

Mutate To change from one form to another.

Native A plant or animal that occurs naturally in Australia.

Nocturnal Active during the night.

Noseleaf A fold of skin on the nose of some bats.

Omnivore An animal that eats plants and animals.

Patagium The thin piece of skin stretching down both sides of some mammals' bodies, which allows them to glide through the air with some control.

Placenta A bag inside the womb of some female mammals that protects young and gives them nutrients before they are born.

Pollinate To transfer the pollen of one flower to another flower. This causes new flowers to grow.

Predator An animal that hunts and eats other animals.

Prehensile Able to grasp and curl around objects (e.g. tails of ringtail possums, pygmy-possums etc.)

Prey Animals that are hunted and eaten by other animals.

Puggle The name given to a young echidna.

Rare Not common.

Rodent Group of gnawing animals, including mice and rats.

Scavenger Any animal that feeds on dead animals it has not killed (e.g. Tasmanian devil).

Species A group of animals that share the same features and can breed together to produce fertile young.

Sperm Special cells the male uses to fertilise the female's eggs during mating.

Spinifex A type of spiky grass found in dry parts of Australia.

Teats The udder of female mammals (except monotremes), where the milk comes out.

Toxins Poisonous products that can cause diseases or death.

Venomous Animals able to use poison, which they inject into their prey.

Vulnerable Can mean either easily hurt or close to extinction.

Widespread Occurring over a large area.

Wingspan The distance between the tips of the wings when they are fully outstretched.

INDEX

A
Aepyprymnus rufescens 33
Antechinus flavipes 8

B
Bilby 16
Brush-tailed phascogale 8
Burramys parvus 22

C
Canis lupus dingo 42–43, 45
Cercartetus nanus 2, 22
Chalinolobus nigrogriseus 36
Common brushtail possum 27, 44
Common ringtail possum 26
Common spotted cucus 26
Common wallaroo 30
Common wombat 20, 48

D
Dactylopsila trivirgata 23
Dasyuroides byrnei 9
Dasyurus hallucatus 11
Dasyurus maculatus 11
Dasyurus viverrinus 10
Dendrolagus lumholtzi 32
Diadem leafnosed-bat 36
Dingo 42–43, 45

E
Eastern barred bandicoot 17
Eastern grey kangaroo 28
Eastern pygmy-possum 2, 22
Eastern quoll 10
Echidna 6–7

F
Fat-tailed dunnart 9

G
Ghost bat 37
Greater stick-nest rat 39
Grey-headed flying-fox 35

H
Hipposideros diadema 36
Hoary wattled bat 36

Honey possum 23
Hydromys chrysogaster 39

K
Koala 3, 18–19, 45
Kowari 9

L
Lasiorhinus krefftii 21
Lasiorhinus latifrons 21
Leporillus conditor 39
Little red flying-fox 35
Lumholtz's tree-kangaroo 32

M
Macroderma gigas 37
Macropus giganteus 28
Macropus parryi 31
Macropus robustus 30
Macropus rufogriseus 31
Macropus rufus 2, 29
Macrotis lagotis 16
Mountain pygmy-possum 22
Myrmecobius fasciatus 14–15

N
Northern hairy-nosed wombat 21
Northern quoll 11
Notomys alexis 38
Numbat 14–15

O
Ornithorhyncus anatinus 4–5

P
Perameles gunnii 17
Petaurus australis 25
Petaurus breviceps 1, 24
Petrogale xanthopus 30
Phascogale tapoatafa 8
Phascolarctos cinereus 3, 18–19, 45
Plains rat 38
Platypus 4–5
Pogonomys mollipilosus 40
Prehensile-tailed rat 40
Pseudocheirus peregrinus 26

Pseudomys australis 38
Pseudomys patrius 41
Pteropus conspicillatus 34
Pteropus poliocephalus 35
Pteropus scapulatus 35

Q
Queensland pebble-mound mouse 41
Quokka 33

R
Red kangaroo 2, 29
Red-necked wallaby 31
Rufous bettong 33

S
Sarcophilus harrisii 12–13
Setonix brachyurus 33
Sminthopsis crassicaudata 9
Southern hairy-nosed wombat 21
Spectacled flying-fox 34
Spilocuscus maculatus 26
Spinifex hopping-mouse 38
Spotted-tailed quoll 11
Striped possum 23
Sugar glider 1, 24

T
Tachyglossus aculeatus 6–7
Tarsipes rostratus 23
Tasmanian devil 12–13
Trichosurus vulpecula 27, 44

V
Vombatus ursinus 20, 48

W
Water rat 39
Whip-tail wallaby 31

Y
Yellow-bellied glider 25
Yellow-footed antechinus 8
Yellow-footed rock-wallaby 30